AF879390

ED'S TRIVIA

TEXTOR'S TANTALIZING TRIVIA
TO
TEASE TORMENTED TECHIES

Copyright 2016 © Waived by Ed Textor
For Personal Educational Uses Only
No Infringements Intended.
Just for Fun by bjbooks
Consultant ~ rickresh

ISBN-13:978-0692668399

TABLE OF CONTENTS

SPIRIT ROVER ~ MARS MISSION

I hadn't realized that Spirit's last "phone home" was two years ago on day 2,210 of its 90-day mission.

The wheels got stuck in soft sand in a crater and attempts to free it failed. Its mission was "re-purposed" to being a stationary radio source to measure fine wobbles in Mars' rotation, which would reveal information about Mars' interior. These measurements aren't possible while moving.

On March 22, 2010, Spirit reported that it was operating normally with remaining systems "green". It's next downlink scheduled for eight days later never came. Engineers anticipated that as the length of the Martian days increased in spring 2010, a higher charge in the batteries would awaken the rover and it would re-establish communications. But Spirit never re-awakened.

Job well-done though Spirit!

I wonder who took Spirit away in that eight days?

ASTRONOMY

In my endeavor to maintain my status as your primary source of useless information, the tidbit I chose to share for this quarter is the whitest object in our entire solar system (other than my legs).

Saturn's moon Enceladus (en SELL' adus) Ἐγκέλαδος, aka, "Trumpeter to Arms" reflects a whopping 93% of the sunlight hitting it because of the pure ice on the surface. If you've used the expression, "pure as the driven snow." this is the perfect example.

Photo #1 shows Enceladus from the Cassini spacecraft with Saturn in the background. Saturn's horizon is visible to the right. The bright lines below it are Saturn's rings nearly edge-on.

[1]

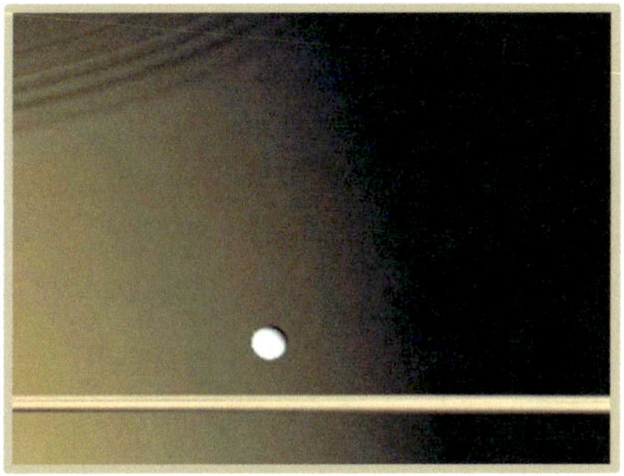

[2]

#2 photo is a Voyager 2 picture taken from about 20,000 miles, and shows Geysers erupting on the surface of Enceladus.

I bet the water would put Crystal Geyser to shame - and I'd be willing to be the first to take a drink!

TITAN ~ SATURN MOON

Now this should bring a tear to your eye. Remember the Pioneer probe, the first to do a mission to slingshot from planet to planet ~ before Voyager? It's on its way out of the solar system, just like Voyager.

A few years ago they'd lost contact with it because it was about at the limit of range for it to keep its dish aimed at Earth. They tried sending a command to it asking for a reply ~ probably something like what we call "pinging." 22 hours later they got a very weak reply. That meant it was eleven "light hours" away from Earth. The command took 11 hours to reach it, and its answer took 11 hours coming back. At the speed of light, that meant it was 7,365,600,000 miles away.

The signal received was 1-billionth the strength of a wristwatch battery. Shortly after, they lost contact and never heard from it again. It has a small nuclear reactor on-board, so it will have warmth and power for thousands of years. But the signal from Earth is now too weak for it to keep its antenna aimed.

It was nice getting a message from it.

Imagine this; Pioneer and the Voyagers someday will be the only evidence that humans existed. Long after the sun swells and swallows Earth in about a half-billion years, those probes will still be moving untouched through space. We have nothing else (yet) that is leaving the solar system. As far as they can calculate, nothing is in their path to hit. So they will be in pristine condition long after our sun is just a white dwarf with some debris orbiting it.

MARS

Last night's sky was so clear and Mars was so bright, straight-up and pink about 9:00. I was tempted to whip-out the old 13-incher and take a look. But Cougar Town was coming on television and my Jack Russell, Holly, wanted me inside to cuddle on the couch. Astronomy isn't a "Russell-approved" activity.
Maybe this weekend.

MARTIAN SEASONS

It's been a while, but this should reinforce my position as your most-reliable source of useless information. Yesterday it became Spring on Mars in the northern hemisphere. Daylight and night were split equally at 12 hours, 18.5 minutes (Martian day is 24 hours, 37 minutes - oddly closer to the human body clock than a day on Earth).

Summer is only 171 days away where temps could soar to highs of 68 degrees. Pack the sunscreen and start planning your vacation!

NEPTUNE 08/08/2011

In my continuing endeavor to be your primary source of useless ~ but sometimes interesting ~ information, this quasi-quarterly's candidate is the planet Neptune. It has been one year on Neptune since its discovery 165 "Earth years" ago. On July 12th, Neptune returned to the exact location in its orbit that it was when discovered by Johann Galle, September 23rd, 1846. On that Fall evening, Johann aimed his telescope and found Neptune in the spot predicted by French astronomer Urbain Leverrier from calculations based-on gravitational perturbations of the other outer planets.

Because of its distance from the sun, Neptune takes 165 of our years to make one orbit. It is 2,795,000,000 miles from the sun, compared to our walking-distance of 93-million miles. Almost 60 Earths could fit inside Neptune it is so big. It's diameter at the equator is 30,775 miles compared to Earth's 7,926. Its surface temperature averages -328, which could be a welcome relief lately.

HAPPY BIRTHDAY NEPTUNE!

JUPITER, VENUS & CONJUNCTIONS

Our view on March 14th. They will appear closer tonight, March 15th when they will be 3-degrees apart. This is only an illusion though. In reality, Venus is nearly 76 million miles from Earth at this time, and Jupiter is 524 million miles from us. So in this picture, Jupiter is about 448 million miles beyond Venus.

Venus being the second planet, is 67 million miles from our Sun (we are number 3 at 93 million). Jupiter, the fifth planet, is 778 million miles from the Sun. Although Venus is nearly a twin of Earth in size, mighty

Jupiter is eleven times bigger at the equator, with a diameter of over 88 thousand miles. (Earth is 8,000 miles). If you could stand on Venus and look toward Earth, we would appear to be the same size as Venus is in these pictures, and we would have a blue tint. You would weigh almost the same, just about ten percent lighter than your Earth weight. But on Jupiter, your weight would be two and a half times what it is on Earth.

Temperature on Venus is about 896 degrees (similar to Fontana). Okay, I'm exaggerating. Jupiter's atmospheric temperature below the clouds hovers around 70. Wind speeds have been clocked at around 300mph (again, similar to Fontana and not much of an exaggeration). We don't know if Jupiter has a "surface" being a gas giant, but if it does, it is speculated that it could be hotter than the surface of the Sun. Jupiter radiates more of its own heat than it receives from the Sun. For this reason, along with strong tidal forces from Jupiter's gravity, icy Europa, a moon of Jupiter, might have liquid oceans under the ice.

Okay ~ fascinate your kids and grandkids while showing them the vastness of our own *little* neighborhood.

MOTION

Continuing in my quasi-periodic attempt to remain your primary and most-reliable source of useless information, I thought I would enlighten you today with **geology stuff**.

The Nepal quake moved Mt. Everest 1.2 inches to the southwest, but did not affect its height of 29,029 feet. It is still a dwarf compared to Mt. Olympus on Mars, the highest volcano in the solar system at 85,000 feet.

Here's more from CNN

[CNN] Mount Everest, the world's highest peak, was moved 3 centimeters (about 1.2 inches) southwest by the magnitude-7.8 earthquake that devastated Nepal in April, Chinese authorities say.

But the April 25 quake, which left more than 8,000 people dead, did not affect the height of the 8,848 meter ~ 29,029-feet~ mountain, according to the report by China's National Administration of Surveying, Mapping and Geo-information.

Chinese state media reported the administration had set a satellite monitoring system on the peak in 2005 to observe the movement of the mountain.

In the decade since, Everest had been moving northeast at a speed of 4 centimeters a year, and had grown by 0.3 centimeters annually, state media reported.

The April earthquake reversed the direction of the mountain, shifting it to roughly where it would have been nine months earlier.

PLUTO

The first clear picture ever of Pluto; about 3.6 billion miles from us but less than a million miles in this picture from the New Horizons spacecraft. The bright area explains the alternating brightening and dimming that made it possible to determine the length of Pluto's day, which is 153 hours, 18 minutes. Pluto has one moon, Charon, that I'm sure we'll see in subsequent pictures from the spacecraft.

One comment posted says that the bright area is Pluto "mooning" astronomer Neil deGrasseTyson for demoting it from planethood.

EARTH STUFF

In my continuing endeavor to remain your primary and most-reliable source of useless information, this quasi-regular ditty (first for 2011) is regarding last week's change of runway designations at Tampa International Airport due to the shift of Earth's magnetic pole.

The magnetic pole is moving eastward and this movement has accelerated recently. On January 13th, the runways will be changed from 18R/36L (180-degrees Right / 360-degrees Left) to 19R/1L (190 / 10).

Granted, this information is useful to some, but this next piece will seal the deal as being the most-useless thing I've told you this year:

The shift of "magnetic north" is not to be confused with the slow shift of "true north". The wobble of Earth on its polar axis means that in 12,000 years, Vega will become our "North Star" instead of Polaris ~ image below. But don't lament. Eventually Polaris will be our North Star again.

Magnetic Field

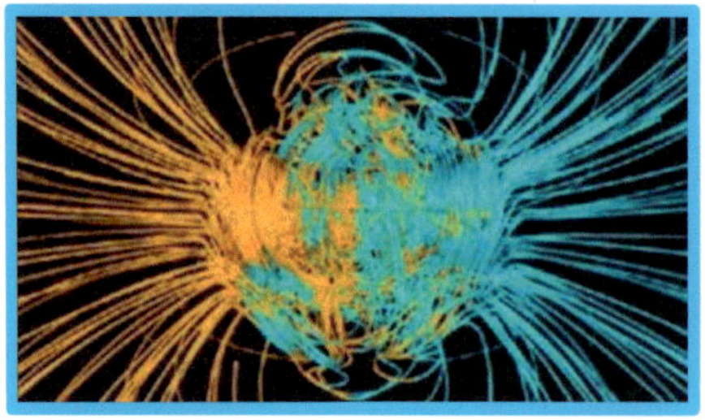

Approximate years for various "north stars." Follow the circle - Vega is around 14,000 AD.

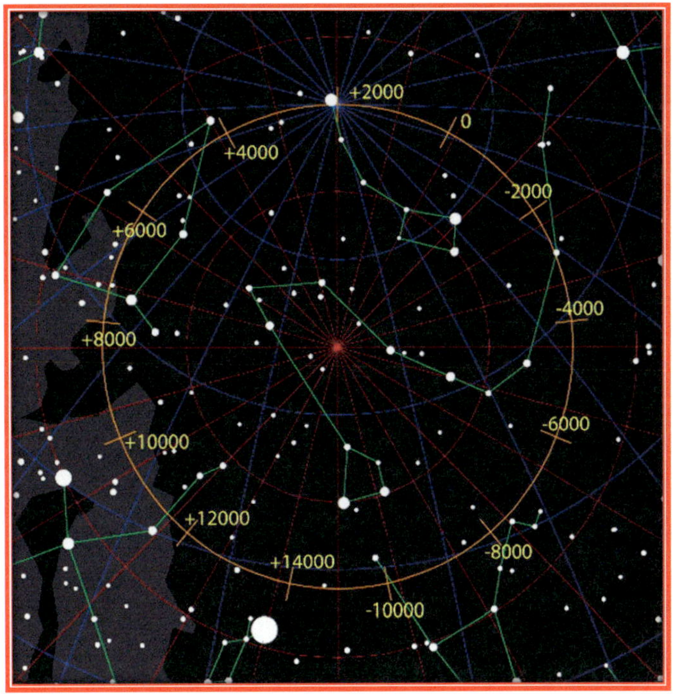

CONSTELLATIONS

This isn't part of my "useless information" series, as I can almost guarantee that at some point in your life, you will be in the position to answer this question. I know in my lifetime, I've been asked no less than twice, and maybe as many as three times. And I'm only 55.

The Pleiades constellation has more than seven stars, Can you identify ***The Seven Sisters***?

The Pleiades (M45), an open star cluster also known as the Seven Sisters, shine brightly just off Taurus the Bull's shoulder, garnering them an extended stay in various cultures' myths.

Here is the answer.

According to mythology, the nine brightest stars include the Seven Sisters, plus their parents Atlas and Pleione a bit to the east.

Asterope1 and 2 are a "binary" system that are so close together, most people can't discern that they are two stars, so they appear as one bright Sister.

But speaking of "binary stars", another little-known "double" is the center star in the handle of the Big Dipper. That is also a binary consisting of two stars, "Mizar and Alcor." You'll likely need binoculars to see these individuals.

It has been very clear this week, so go outside tonight ~ after submitting your EMACS ~ and see how many you can count. Then dazzle your friends with this little ditty of info!

SIDEREAL TIME

Because of Earth's forward motion in its orbit around the sun, a "sidereal day" is about four minutes shorter than a "solar day". In other words, if you choose any star (star, not planet like Venus but a real star like Sirius) and note its position at some time in the evening, say 9:00 PM, the next night it will be at that same spot at 8:56 PM. The next night at 8:52, etc. This is because in addition to our rotation, we are also moving at about 12,000 miles per hour around the sun and that impacts our viewing angles.

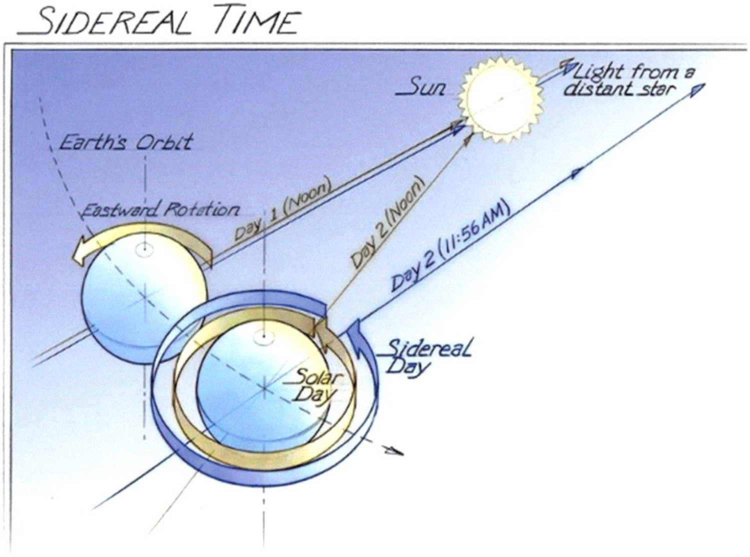

TIMESTUDY

Timestudy is what these guys do when trying to explain the size of the universe as compared to the age. They're all so he11-bent on the impossibility of matter exceeding the speed of light, yet in order to explain the size of the universe, everything must have been exceeding the speed of light for 3-tenths of a second after the big bang. Since they say that's not possible, we must have 10-dimensions. Yeah, that's a stretch. I was content to stop at the Fifth Dimension. That was great music.

But there's a new theory now. "They" say that sure, matter cannot exceed the speed of light through space. But maybe space itself can exceed the speed of light, dragging everything along with it, thus not violating "their" perceived laws of physics.

So I guess we're full-circle back to, "the universe must have been expanding faster than light for the first three-tenths of a second after the big bang".

How fast will we be moving when we get sucked back into the "big crunch"?

MOON

If you are younger than 38, this will be your first Christmas Day full moon. We haven't had one since 1977 and we won't have another until 2034.

The last full moon of the year is called the "Full Cold Moon" because it falls in Winter. It will peak at 3:11 AM Pacific Time.

Set your clocks to wake-up the kids and grandkids for this; you might even catch a glimpse of Santa and his reindeer passing in-front!

LAST NIGHT'S FACEBOOK POST

Ed Textor added 4 photos.

NASA confirms Voyager 1 left the solar system about a year ago. This is an incredible accomplishment for our civilization. Think about it; our generation is the first to send a spacecraft out of the solar system into interstellar space. Voyager 1 is now beyond the gravitational influence of our sun, never to return. Ever.

Long after the Earth is gone and the sun has burned-out, Voyager 1 will still be traveling in darkness, carrying pictures of us, bits of our history, and messages of greetings from the people of planet Earth in 60 languages with samples of our music on a gold record entitled "The Sounds of Earth". There are diagrams of the human body, and a chart showing our exact location in the cosmos. Binary math is included to help translate the written messages. Any intelligent species will know binary math.

Voyager 2 will be the next to enter interstellar space, followed in a few years by Pioneers 1 and 2. Four little spacecraft that left Earth back in the seventies, and had already provided so much information about our own neighborhood of planets, are now making the ultimate leap into history.

Damn we're good!

http://news.nationalgeographic.com/news/2013/13/130911-voyager-interstellar-solar-system-nasa-science-space/

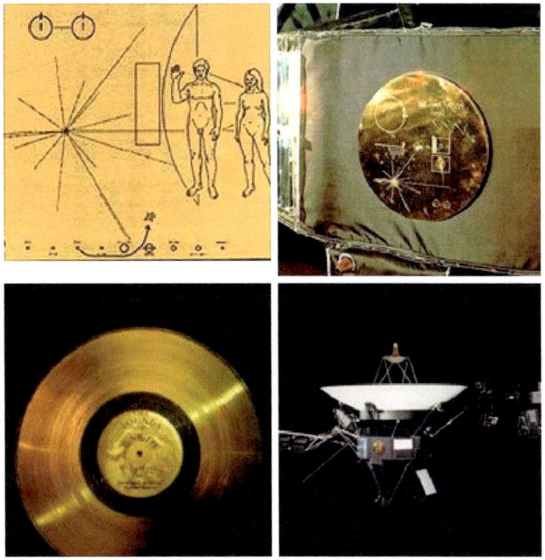

THE SUN ~ aka ~ "SOL"

Okay, since Bones went and asked a question involving the "A" word (astrology) the thought of which with me, triggers involuntary shuddering, chills, and feelings of "nails dragging on a chalk board" or what we old people used to watch teachers write on in school. I now feel compelled to give you all a new ditty of information. Again, because I want to remain your most reliable source of useless information.

"Sol." our star, rotates once every 25 days. That means its one-day is equal to 25 of ours.

The star Altair which is seventeen light years away [meaning if you could travel at 669,600,000 miles per hour it would take you seventeen years to get there], rotates once every 9 hours. It is rotating so fast it actually bulges at the equator and is 20% wider at the equator than at its poles.

Altair is about 1.8 times bigger than Sol, and is a Spectral Type A7 which means it is hotter and a brighter-white than Sol which is a class G2. So if you were on a planet orbiting Altair, you would want your planet a good distance farther than the 93-million miles we are from Sol. Otherwise your entire planet would look like Yucca Valley (sorry Mitzi)!

Spectral types ranging in this order from hot blue-white to red dwarf are: OBAFGKM. F, G, and K would be the most hospitable. You can see we are right in the middle of the ideal types.

A little phrase is used to remember the correct order of Spectral Types, and is, "Oh Be A Fine Girl Kiss Me". There, now you'll never forget.

TV TRIVIA

Altair was the star chosen for the home planet in the movie "Forbidden Planet."

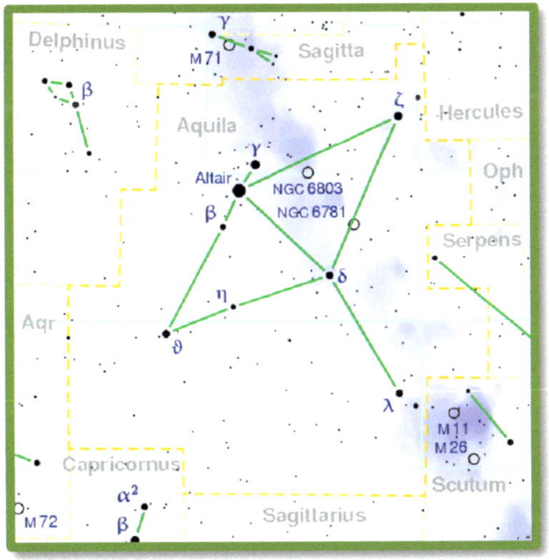

You asked for it McCoy!

TIME TRAVEL

Quote: "You can't travel into the future because it hasn't happened yet. You can't travel back because it's not in our history that it happened."

So I hear astrophysicists talking about the Big Bang and such. Then they say things like, "Well in order for the universe to be as it is today, that means in the first microseconds after the Big Bang matter would

have to be traveling faster than light. But because matter can't travel faster than light, there had to be ten dimensions! That's where they lose me in absurdity. Like they were there and witnessed the ten dimensions?

Ed says, "Heck, maybe matter *can* go faster than light if it's kicked hard enough ~ ~." And since you need to be able to move at light speed to alter time ~ ~ .

But then they talk about the old paradox of 'going back in time and killing your grandfather ~ ~." Recently they're saying, "You can do that but you won't alter your own history. You will cause a fork in the river of time and create a parallel timeline where you'll never exist.

Then they say there is no limit to parallel timelines and there are already parallel timelines where you don't exist. Another where I don't exist or maybe one where I'm a good-looking Olympic athlete?

DOG DAYS OF SUMMER

The term "dog days of summer" are named after Sirius (the dog star). Now, you say, "But Sirius isn't up in the summer months..." and you are right. Sirius is up through the coldest months of our year. However, the ancient Egyptians didn't have a grasp of the distance of stars from us. They noted that it is the brightest star in the northern hemisphere, therefore, must be the hottest. So, it's appearance in the late winter months must be what triggers our warmer weather months later. So they did have an idea that it was far away, and it wasn't a bad analysis actually. But at eight light years, it is too far to have any influence on our weather.

Al Gore's hot air has more effect.

'73 IS THE BEST NUMBER

73 is the 21st prime number.
Its mirror [37] is the 12th and its mirror [21] is the product of multiplying 7 and 3.
In binary, 73 is a palindrome, 1001001 which backwards is 1001001."

Another ditty of information about 73 not mentioned in the show ~ ~ The square root of 73 is 8.544.
The sum of which (8+5+4+4) equals the product of 7 and 3; 21.

This episode is the 73rd episode in the Big Bang Theory series.

And of course - *Really cool people graduated from high school in 73.*

INTELLECTUAL EGO

To a 10 year old boy who bought Binoculars for looking at Stars.

"Dear binocular user,

I am the Chief Senior Astronomer at the Mauna Loa observatory 14,000 feet above the Hawaiian Islands. Our telescopes can see farther into the universe than anything else on planet earth, second only to the Hubble Space Telescope in orbit 380 miles above the surface, occasionally serviced by shuttle astronauts at least until we run out of shuttles and/or astronauts.

I have noticed recently that you consistently are trying to use binoculars to see objects fainter than the 6th magnitude naked-eye limit. As Chief Senior Astronomer with a Masters in Astronomical Observation, it is my duty to inform you that although it is possible to see faint objects with binoculars, you would be much more successful if you mounted them on an equatorial mount tripod with a counter-balanced electronic clock drive synchronized with sidereal time. However, it is unlikely you will be able to see stars in the spectral-class K or M, but you should have no problem finding the brighter spectral classes O, B, A, F, and G. Of course, I'm sure you are already aware that our sun is a spectral-class G-IV, not too different from our nearest stellar neighbor Alpha Centauri which is a class G-V.

If you are unable to find the object you wish to observe, check your right-ascension and declination coordinates after verifying that you have correctly oriented the tripod to true-north. An easy way to do this is to locate Polaris via the Sky Catalogue 2000.0 coordinates of RA=0 and D=0.3., and orient your binoculars using your RA & D dials.

Ed Textor, Masters in Astronomical Observing, Chief Senior Astronomer, Mauna Loa, Hawaii.

COMPUTERS

Last night while browsing available Linux applications, I came-across this app with a summary that I'm sure is enlightening to someone. But after reading it three or four times, I was still totally clueless on what this thing actually does. I'll buy lunch for the Tech who can tell me what this does!

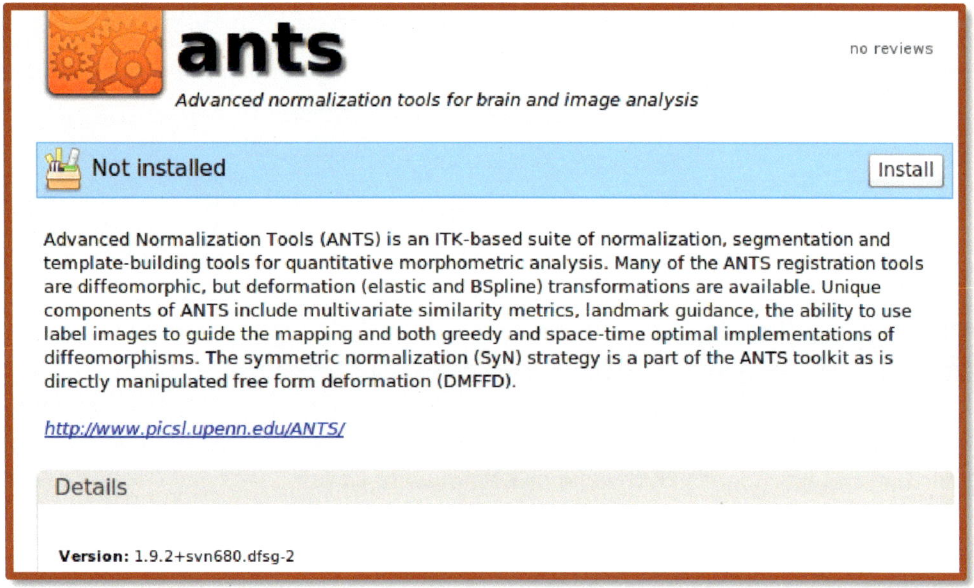

USING BINOCULARS

In deep thought while pondering the use of binoculars instead if deploying Disk Encryption Software:

If I've never told you the nice trick for comfortable binocular viewing ~ so you are not standing with your arms over your head trying to hold still, sit at a table and lay a mirror on the table. Then you can rest your elbows on the table and be looking down at the reflection. Tip the mirror slightly to view different parts of the sky.

So have I tried this? No! But it's a great idea.

CHAT ROOM

Speaking of Finland though; my first few weeks of having any internet access many years ago was with AOL. I was in a chat room about an upcoming eclipse; can't remember which one.

A guy in the chat room mentioned he would have his telescope outside with a CCD camera on it feeding to his big-screen TV for him and his friends that were coming over to watch.

I asked him where he was. He said, "Finland."

That response floored me - and I realized many fantastic things were ahead for us in technology. The idea that I could be communicating with a person in Finland for no charge (other than the flat AOL charge) on the other side of the world, about a celestial event. I was awestruck.

DATA CENTERS

The 10th largest river, by water volume, in California is the discharge pipe from the El Segundo sewage treatment plant. Coincidentally, Southern California's Welfare data center is in El Segundo.

WINDOWS, DOS AND SCRIPTS

The batch files use the CACLS command to set permissions. Server 2008 now has 'power shell' instead of just the 'command' (DOS) window. I put in CACLS /? to look at the help, and got a message that "CACLS is 'deprecated'." Now that sounded ominous. It then said to use iCACLS, which seems to have more command options.

dep·re·cate
/ˈdɛprɪˌkeɪt/ Show Spelled [dep-ri-keyt] Show IPA
verb (used with object), -cat·ed, -cat·ing.

Definitions ~
1. to express earnest disapproval of.
2. to urge reasons against; protest against (a scheme, purpose, etc.).
3. to depreciate; belittle.
4. Archaic ~ to pray for deliverance from.

I perform the final definition every time I log on to Windows.

TECHNICAL SUPPORT

I was dragged ..... I mean, "invited" into the ITSD family in 2003. Before that I was with JESD since '94. Started with the county in 90. I worked as a Supervisor for Field Technicians in a large IT environment as well as Help Desk Staff Supervisor. I had always heard of the old "Tech Support Idiot Calls" that usually degrade end-users who aren't, shall we say "Advanced Users" of computers. I honestly thought they were joking. Then one day, one of my Techs sends this Support Call outline:

~ ~ ~ ~

HelpDesk Call 357XXX

Summary: User unable to view new e-mails she's creating.

8/10/2015 3:37:46 PM D2853

Problem: User unable to view new e-mails she's creating.

History:

Troubleshooting: Remoted to user's computer; had user turn on right screen; she can see e-mail.

Yes, Virginia, they are REAL!

THINGS THAT MAKE YOU GO "HUH?"

English Language

You've heard me talk about inconsistencies in our language, the shared "too" that means '*also*' as well as '*excessive*' when we should have a fourth ~ '*tew*' for one of those.

And where was that fabled "w as a vowel" in our language that we heard about in the sixties ~ my daughter never heard that in the 80s ~ but in the teachers had no examples beyond "helping a real vowel by altering the sound," (as in aw)? I don't count 'helping' as being the real thing.

Real w-vowel examples: Crwth - a Celtic musical instrument pronounced 'cruuth' (literally a double-u) and cwm - a valley pronounced "koom" (credit to Mitzi for that one - actually Welsh but is in English dictionaries.

So another question of mine is, why when we have, "thirteen, fourteen, etc.", do we use "eleven and twelve" instead of "firsteen, secondteen"?

And I wonder why I'm an insomniac.

Footnote:
The counting thing also is found in Spanish --- one would think it should be "dos y uno" (reference Baenti-uno, etc.) ~ hmmm, Dos y Dos ~ dosey doe ~ Shall we dance?

I never snore, either.

UP ~

***A**n English language word that can be a noun, verb, adj., adv., or a preposition.*

This two-letter word in English has more meanings than any other two-letter word, and that word is *'UP.'* It is listed in the dictionary as an [adv], [prep], [adj], [n] or [v].

It's easy to understand *UP*, meaning toward the sky or at the top of the list, but when we awaken in the morning, why do we wake *UP*?

At a meeting, why does a topic come *UP*? Why do we speak *UP*, and why are the officers *UP* for election and why is it *UP* to the secretary to write *UP* a report? We call *UP* our friends, brighten *UP* a room, polish *UP* the silver, warm *UP* the leftovers and clean *UP* the kitchen. We lock *UP* the house and fix *UP* the old car.

At other times, this little word has real special meaning. People stir *UP* trouble, line *UP* for tickets, work *UP* an appetite, and think *UP* excuses.

To be dressed is one thing but to be dressed *UP* is special.

And this *UP* is confusing: A drain must be opened *UP* because it is stopped *UP*.

We open *UP* a store in the morning but we close it *UP* at night.

We seem to be pretty mixed *UP* about *UP!*

To be knowledgeable about the proper uses of *UP*, look *UP* the word *UP* in the dictionary. In a desk-sized dictionary, it takes *UP* almost 1/4 of the page and can add *UP* to about thirty definitions. If you are *UP* to it, you might try building *UP* a list of the many ways *UP* is used. It will take *UP* a lot of your time, but if you don't give *UP*, you may wind *UP* with a hundred or more.

When it threatens to rain, we say it is clouding *UP*. When the sun comes out, we say it is clearing UP. When it rains, it soaks *UP* the earth. When it does not rain for awhile, things dry *UP*. One could go on and on, but I'll wrap it *UP*, for now ~ my time is *UP!*

Oh ~ one more thing: What is the first thing you do in the morning and the last thing you do at night?

U - P

Vowels

Talking with my daughter years ago about the vowels; when I was little we were taught: A, E, I, O, U, and sometimes W and Y. My daughter had never heard about the W. 'Y' is easy. In a word like *Tryst,* it is the only vowel. When you research the net about the W, you'll find comments like, "it's a vowel "helper," meaning it changes the sound of a vowel. Example, "Aw".

My teachers were never able to give an example. I didn't think "vowel helper" was good enough. I think the original intent of listing it means it exists in some words as the "only vowel," just like Y in tryst.

There are some Welsh words, from which our language has some roots that use W as the only vowel.

And the sound is literally a "double-u." Picture uu which looks a bit like a w.

On sharing this information in my "useless" series, Mitzi Cruz found one, 'Cwm', a Welsh word for 'valley' that was anglicized to "Coombe."

So this solved my age-old question of "just where are these sometimes Ws?

Crwth is a Welsh musical instrument, and the pronunciation rhymes with *truth*. Some might say, "But it's Welsh, not English." But some of our dictionaries include it, so I say, it counts as English too.

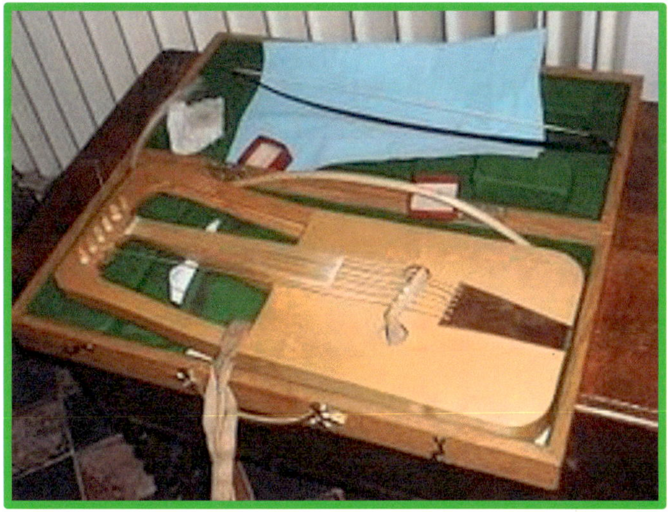

A Crwth In Case

Pastoral Comments

I should have trusted Probie. Which reminds me of something a pastor said once. People spend too much time saying, "I should've". He said, "you have to stop should-ing on yourself".

GENERALITIES

Did you know cigarettes aren't hot enough to ignite natural gas? Not that I'd want to test that. A fire captain once told me the scene in the movie "Earthquake" where the guy with a cigarette runs back into the house and it explodes, was bogus.

On the subject of Hollywood; my other peeve is, unlike in the movies, only one sprinkler pops if you hold a flame to it. So every show you see where someone flames a fire sprinkler and the whole building goes off, bogus. But they love that.

And then there's the old, kick the extension cord in the swimming pool and everyone gets electrocuted. Nope.

Okay, we're on a roll now.

Noise in space. How many space movies do you see with rocket and explosion sounds. No sound in a vacuum. And guess what! When you run-out of fuel in a spaceship, you don't coast to a stop.

And sorry Bruce Willis; the space shuttle could never have gotten to the moon. Especially in the ninety seconds it took you to get there after you spilled fuel all over the cargo bay when refueling ~ but where were you putting the fuel when your tank was on the bottom of the Atlantic by then?

And please turn-off the face lights inside the helmets. You can't see out into the darkness when your face is lit-up like Vegas.

Okay, I'm out. ☺

Giant Concrete Arrows Pointing Your Way Across America

Every so often, usually in the vast deserts of the American Southwest, a hiker or a backpacker will run across something puzzling; such as, a large concrete arrow, as much as seventy feet in length, sitting in the middle of scrub-covered nowhere.

What are these giant arrows? Some kind of surveying mark? Landing beacons for flying saucers? Earth's turn signals?

No-o-o-o, they are directionals for ~ ~ ~ ~

THE TRANSCONTINENTAL AIR MAIL ROUTE

On August 20, 1920, the United States opened its first coast-to-coast airmail delivery route, just 60 years after the Pony Express closed up shop.

There were no good aviation charts in those days, so pilots had to eyeball their way across the country using landmarks. This meant that flying in bad weather was difficult, and night flying was just about impossible.

The Postal Service solved the problem with the world's first ground-based civilian navigation system: a series of lit beacons that would extend from New York to San Francisco. Every ten miles pilots would pass a bright yellow concrete arrow. Each arrow would be surmounted by a 51-foot steel tower and lit by a million-candlepower rotating beacon. A generator shed at the tail of each arrow powered the beacon.

Now mail could get from the Atlantic to the Pacific not in a matter of weeks, but in just 30 hours or so.

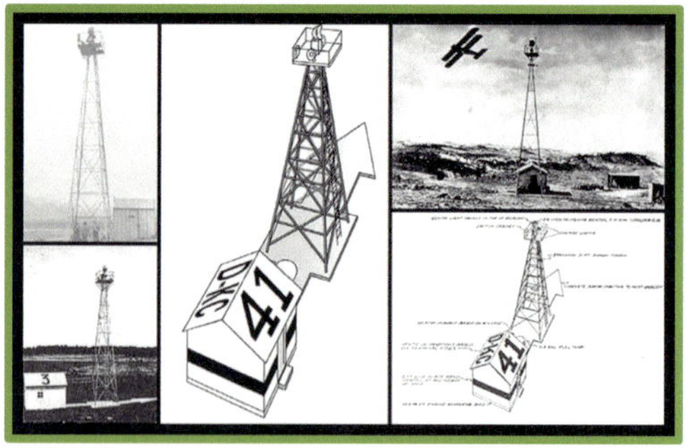

Even the dumbest of air mail pilots, it seems, could follow a series of bright yellow arrows straight out of a Tex Avery cartoon. By 1924, just a year after Congress funded it, the line of giant concrete markers stretched from Rock Springs, Wyoming to Cleveland, Ohio. The next summer, it reached all the way to New York, and by 1929 it spanned the continent uninterrupted, the envy of postal systems worldwide.

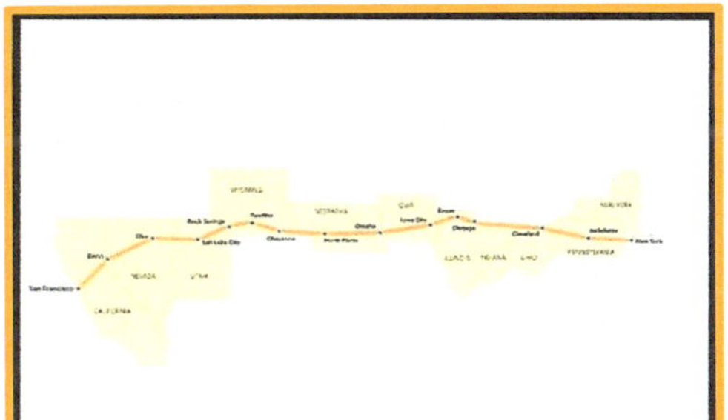

A re-creation of a 1920s map showing the route of airmail planes; the dots are intermediate stops along the course.

Radio and radar are, of course, infinitely less cool than a concrete Yellow Brick Road from sea to shining sea, but I think we all know how this story ends. New advances in communication and navigation technology made the big arrows obsolete, and the Commerce Department decommissioned the beacons in the 1940s. The steel towers were torn down and went to the war effort.

But the hundreds of arrows remain. Their yellow paint is gone, their concrete cracks a little more with every winter frost, and no one crosses their path much, except coyotes and tumbleweeds.

But they're still out there!

VERY INTERESTING FACTS ! !

One Year Ago Today

Flight home from Aberdeen Washington after assessing our flooded house up there and meeting with the insurance adjuster, contractor, bleaching the wet floors, mopping-out mud. Didn't know what was ahead for the claims and work, but the contractor already had hot air blowing under the house after ripping-down the floor insulation. All worked-out just fine about six weeks later and the house looked better than before on the inside. Our bleaching and his fast-drying response prevented any mold from even getting started. He loved my idea of cutting 1-1/2-inch holes through the baseboards every 16-inches in all rooms; that accelerated the drying in the walls. We had high-capacity fans running inside and he had the blowers under the house for a week. The humidity in the walls dropped to an acceptable level within three days of us leaving.

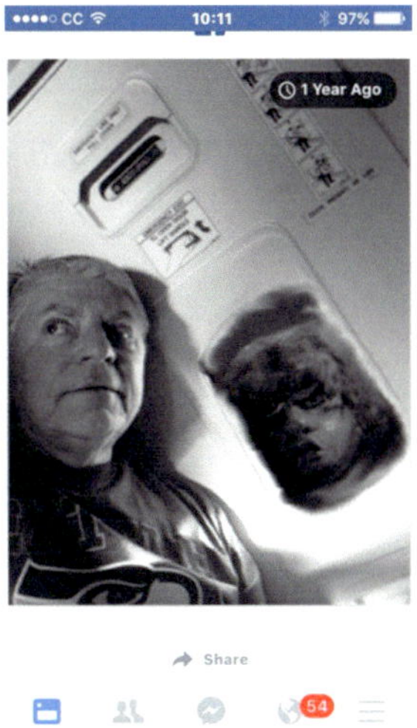

Life and Death of a Penguin

Dead Penguins - I never knew this! Did you ever wonder why there are no dead penguins on the ice in Antarctica? Where do they go? Wonder no more! It is a known fact that the penguin is a very ritualistic bird which lives an extremely ordered and complex life. The penguin is very committed to its family and will mate for life, as well as maintain a form of compassionate contact with its offspring throughout its life. You really didn't believe that I knew anything about penguins, did you?

If a penguin is found dead on the ice surface, other family members and social circle are known to dig holes in the ice using their vestigial wings and beaks, until the hole is deep enough for the deceased bird to be rolled into and buried. The male penguins gather in a circle around the fresh grave and sing:

"Freeze a jolly good fellow,
Freeze a jolly good fellow."

Oh, quit your whining, I fell for it too!

COMMUNICATIONS

Bluetooth Technology

Something that will make your friends go "hmmm" and contribute to maintaining my status as your primary source of useless information ~

Ericsson originally called their technology 'Bluetooth' after Harald Bluetooth, who was king of Denmark between 940 and 981. During his rule, Denmark and Norway were Christianized and united, so Ericsson used the analogy that he "allowed greater communication between people" when naming their wireless communication protocol. ~ Richard Thompson, Allerod, Denmark

Ham Radio Theory

Haven't you always wondered where the term "HAM radio" came from? As your consistently-reliable source of useless information, I found the following ~

Nobody is really sure, not even the authorities and old-timers in the hobby itself. The formal designation for that service is "Amateur Radio".

One theory is that H.A.M. is an acronym paying homage to three great radio pioneers: Heinrich Hertz (demonstrated existence of electromagnetic waves); Edwin Armstrong (invented modern FM radio), and Guglielmo Marconi (developed wireless telegraphy).

Another theory is when the first amateur radio operators where using CW (Morse Code), that they tended to have a '**heavy**' (hard) hand, and was called being "ham fisted". This tends to be the most common theory.

I don't know about that second one. I think I'm favoring the first theory by about 70%; and throwing a token 30% to the second one.

And a Big 10-4, Good Buddy!

ENDNOTES

Latin Translations - http://www.inrebus.com/index.php

Bene legere saecla vincere - To read well is to master the ages. (Professor Isaac Flagg).

Omne tulit punctum qui miscuit utile dulci - [He] has gained every point who has combined [the] useful *and useless* [with the] agreeable.

Absque sudore et labore nullum opus perfectum est - Without sweat and toil no work is made perfect. Also - From a clear spring clear water flows.

Ab honesto virum bonum nihil deterret - Nothing deters a good man from the performance of his duties.

Amicorums optimi in aeternum – Friend Forever

Aequam memento rebus in arduis servare mentem - Remember when life's path is steep to keep your mind even. (Horace)

Aequam memento rebus in arduis servare mentem, non secus in bonis ab insolenti temperatam laetitia - In arduous circumstances remember to preserve calmness of soul.

Gunsmoke 1955-6, Encore Western reruns 2015

1874 Dodge City, Kansas
Matt to Telegrapher: "Can you tell me who sent this telegram?"
Telegrapher: "Sure, Marshall, Any operator on the line can be identified by his "fist."

Made in the USA
Las Vegas, NV
09 November 2026

57757232R00040